ABOUT THE AUTHOR

The author has a master's degree in organic horticulture from the University of Agriculture and Horticulture, Iasi, Romania. He is passionate about nature and the impact of plants and trees on our lives. He has a variety of interests, including writing short fiction novels, poetry and theology. He is a published author in Romania. This is the first of his books to be published in the UK.

Thank you!
Rock Bodies
2023

RADU BOBICĂ

The Magic of Walnut

AUSTIN MACAULEY PUBLISHERS™

LONDON • CAMBRIDGE • NEW YORK • SHARJAH

A CIP catalogue record for this title is available from the British Library.

ISBN 9781398474840 (Paperback)
ISBN 9781398474857 (ePub e-book)

www.austinmacauley.com

First Published 2023
Austin Macauley Publishers Ltd®
1 Canada Square
Canary Wharf
London
E14 5AA

All my thoughts and love to John for the help and support. Thank you for pushing me to do this till the end, and for believing in me. I could not do this without you.

Many thanks to Stephanie, Saul, Ramona, Jessica and Rohit, for all the support offered. Love you all.

Emma and Auryn,
Nothing in my life is more important than you two.

CONTENTS

Chapter 1

INTRODUCTION

A N OLD Romanian saying states, "Nothing grows in the shade of the walnut tree." The truth in this ancestral proverb has two meanings. Figuratively, the walnut is associated with a person who has created a name for themselves, a reputation, or status. No one can develop in the shadow of a strong person. Due to the properties of the walnut sprout or tree, other plants face challenges, such as difficulty growing under certain conditions. There are only a few plants, such as gooseberries, raspberries, rhododendron, magnolia and azalea, that can grow under a walnut tree. That is because they like acidic soil and can tolerate the toxic substances released by the walnut. That is not all; walnut leaves release a large amount of iodine and juglone, toxic substances that can reach the ground through rainfall. This is one of the main reasons for the poorly developed or non-existent vegetation under the walnut tree. Walnut shade is not suitable for insects either, and it is understandable that staying under the shade of a walnut tree or leaf is not always recommended. That is because during the autumn, there is a risk of black mould appearing on the remaining nuts hanging on the tree. This is possible, especially if the humidity is over 75 per cent. Staying under a walnut tree for a long time increases the risk of inhaling the mould spores. In spring, green walnuts and leaves may leave stems on your skin, which can cause irritations.

I have always considered walnut trees to be the most valuable trees we have with us. But as valuable as it is, little is known about its value, so we miss out on its true potential. God has blessed every part of this tree, from the wood, which is considered an aristocrat among other types of wood, to the leaves, with their pharmaceutical properties, and the fruit, which, unlike the fruits of other tree species, constitute a complete and concentrated food source. So, all walnut components can be used and are used for the benefit of humans.

It should be noted that the walnut tree is among the few trees that absorb heavy metals from the atmosphere. This must not be

Green walnuts

omitted when we want to harvest leaves, fruits or walnut flowers for use in treatments or consumption. Trees or plants that have curative properties can very easily retain polluting substances in their attempts to cleanse the air. Strong roots give the walnut a significant role in protecting the soil from erosion and landslides, being successfully used in the creation of flood protection barriers.

To have a walnut in the garden is a gift, a marvel, especially if we consider that only seven per cent of the earth's surface is suitable for the cultivation of this tree.[1] So, if you're lucky and the walnut tree grows in the soil around you, then care should be taken when protecting the walnut tree, allowing future generations to enjoy and benefit from its properties. A walnut legacy can be left in some countries like Iran, Afghanistan, Kazakhstan, Turkmenistan, etc. There is an old tradition that a walnut tree is planted as dowry when a child is born, as its lifespan is roughly three centuries. It also provides a future source of income for the child. In the fourteenth century, the main units of exchange between some early Muslim countries and West Africa were gold, ivory and walnuts. This helped a lot to develop the economies of those regions.

The walnut family is the Juglans and has over 20 varieties, the most common being Juglans Directed and Juglans Nigra. In what follows, we are going to explore the most common varieties of Juglans.

1   Statistic from wikipedia.org.

Chapter 2

A SHORT HISTORY AND DESCRIPTION OF THE WALNUT

T HE WALNUT is native to central and southwest Asia. Specialists believe that it has survived in isolation in these areas since the Ice Age. Around the same time, the walnut has been a survivor throughout several depressions, for example in

Walnut in tree crown

the Balkan Peninsula and the Iberian Peninsula. Today, we can still find the walnut in the wild, in the forest massifs of eastern Turkey, India and China, as well as in several areas of Peru and Brazil.

Walnuts spread very quickly in temperate and Mediterranean biomes, both as a spontaneous flora and in crops. Before the Christian era, there were tribes around the world that cultivated walnuts in countries such as China, Japan and India. Today, the most intensive nut crops are in France, with over 20,000 hectares in the Perigord area, and in California, U.S., with over 100,000 hectares of orchards with selected varieties.

With cultivation, the walnut has extended far beyond its natural range, currently being found in almost all countries with temperate climates and can increase to altitudes of 800 metres. The northern limit of its cultural expansion around the globe is imposed by the annual isotherm of +5°C.

Scientific classification of the walnut:

Kingdom – *Plantae*
Division /Phylum – *Magnoliophyta*
Class – *Magnoliate/Dicotyledons*
Order – *Juglandales*
Family – *Juglandaceae*
Genus – *Juglans*
Species – *J. Directed*

It was the botanist Carl von Linné, or Linnaeus (1707–1778), who "signed" the official birth certificate of the walnut tree by devising its scientific name.

Von Linné used the appellation *Iovis glans* – hence, *Juglans*. The common walnut tree had the epithet "royal" added to its name, *Juglans regia*. If we go back to our Latin ancestors, the name *Iovis glans* would translate as "Jupiter's acorn" – the name of the God of Heaven. It is said the Romans were the ones who, at weddings and holidays, threw nuts in the way of thanksgiving as a symbol of abundance and well-being. Although, in all probability, due to the harsh blows received by those who came into contact with a walnut thrown with force, this custom has now been replaced by throwing wheat or rice. In another part of the world, the Caucasus walnut is considered a holy tree and can survive for over 400 years. In Hindu rituals, we find walnut as part of the holy festivals and weddings. There is extensive use along the Kashmir Valley and different parts of northwestern India.[2]

In many countries in Asia and Eastern Europe, walnut fruits are used as gifts exchanged between friends and family members. They are also used as offerings given to Gods or celebrants, who can visit a family home on different celebratory occasions.

In Romania, the tree is protected by law, and cutting it, even a scraped tree on private property, is punishable by a rather stinging fine. A walnut can be cut under certain conditions (climate or other accidents), but even so, the owner of the land is required by law to plant three walnuts in exchange for the cut one.

Returning to the name of the walnut, we notice that in the Roman Empire, the name derives from the Latin *nux*; to Romanians, this has become *nuc*; to Italians, *noche*; to Spanish, *nogal*. On the outskirts of the Roman Empire and its Germanic and Slavic neighbours, however, the name of the walnut comes from the Old German word *Walch*: in German, *Walnuss*; in English, *walnut*; in Danish, *walnod*. This word *walch* is what the Germanic peoples used to refer to foreigners: Romans, but also Romanians – *Valahs*. In Slavonic languages, the walnut is called *valah*: to Ukrainians, горіх волоський; to Czechs, *orech vlassky*; to Poles, *orzech wloski*.[3]

The walnut is a vigorous tree that can reach up to 30 metres high. Its trunk is thick, the bark smooth, and it is silver-grey

---

2   Vahdati, 2014.
3   Information from Ion Comanic.

in colour. The walnut crown is wide and rich, and has strong branches, providing shade and coolness on warm days.

Walnut trees, whitewashed to reduce exposure to the sun

With such a peculiarity as the crown, it is no wonder that its roots are also very strong. They can easily invade the area they need for better fixation (the nut will always stay above flood spills). Like any tree, the roots of the walnut tree lie beyond its crown. The leaves are large, composed of 5–9 elliptical follicles, with whole edges and no hair.

Walnut is a manioc (monoecious) unisex species and forms two types of flowers (which are on the same tree): male flowers and female flowers. Even so, some of Central Asia's guardians are convinced that the walnut tree never blooms. According to a Japanese proverb, "Let the one who sees the walnut flower die."

The walnut fruit is a spherical drupe[4] with 4–5cm in diameter and a fleshy coating that breaks apart in maturity, thus

4   See definition on page 24.

releasing the ovoid and irregularly crossed walnut. Inside the nut is a real energy bomb – the seed, the walnut core, which is among the most calorie-rich and nutritionally complete natural foods.

Walnut fruit

Chapter 3

WALNUT CULTURE

WALNUTS GROW spontaneously in many areas, but that does not mean that scraped trees can meet consumer requirements. There are farmers who go to the trouble of setting up walnut plantations. Such farmers are not only to be commended but should also be properly supported by local

Young walnut trees: they grow spontaneously

and government authorities. The move of a farmer who wants to establish a walnut crop must be supported, primarily because he will not get an immediate gain from the newly established crop, but the expenditure will be commensurate. To buy a single medium walnut tree can cost up to £212. Sometimes, farmers purchase kits that support the healthy growth of a walnut tree.

The walnut tree is not demanding once planted; the soil and climate largely protect the tree from diseases and pests. However, the greatest disadvantage is the time taken for the walnuts to fully develop. It can take as much as three to four years after planting for the tree to fully grow, depending on the walnut chosen. However, there are some species of walnut that give their first fruit or walnut more than 10 years after planting. Therefore, sometimes there is no immediate profit.

The walnut tree is grown by planting a walnut in the ground that will germinate in the nurseries, and then (depending on the variety) the sapling is planted in the field after three to four years. Each walnut variety requires a certain planting scheme, but on average, one hectare of land can be used to plant about 140 to 160 saplings. It is recommended that the planting of nuts be completed in the autumn, when the plants stagnate from the vegetation. Therefore, this alters the time of planting and ultimately begins the vegetative cycle quicker. Of course, you can also choose to plant saplings in the spring. When planting, we must consider that nuts prefer a well-drained soil with a neutral pH.

Regardless of the period during which the saplings were planted, the crown-forming cuts begin early in the spring and apply throughout the growing period. The purpose of these cuts is to form an airy crown (which will naturally protect the tree from diseases and pests and allow better fruit formation), an easy crown (light conditions and rate of growth and formation of the aerial organs of the plant), and in particular a garnished crown with fruit branches, as soon as possible. There are various forms of crown formation, from an improved vessel to a staged pyramid, etc. However, there are researchers who argue that at trunks taller than 1–2 metres, the fruit is delayed by one year, for every 20 cm more.[5] Therefore, upon cutting walnut trees, their growth needs to be considered to prevent a late fructification cycle.

Unlike the rest of the fruit trees, the walnut does not have many natural enemies (diseases and harm). The high levels of iodine released by the leaves of the walnut tree cause insects and pests to bypass it. That does not mean it is immune. It also has its predators: these have to be combated through human interventions such as proper cutting and respecting planting distances. If these are not effective, environmentally-friendly treatments such as Bordeaux juice (a mixture of copper sulphate and quicklime) can be used. When used properly and in the recommended concentrations, natural products are effective for protecting walnuts.

The use of chemical treatments should only be done in cases of urgency and after thorough analysis. The efforts to combat diseases and pests aim to maintain the health of the leaf apparatus in order to synthesize maximum amounts of substances that ensure the tree's vigorous growth, especially in young plantations.[6]

5   Dr Gelu Corneanu PhD and Dr Margareta Corneanu PhD.

6   Information from the works of Dr Gelu Corneanu PhD and Dr Margareta Corneanu PhD.

Chapter 4

WALNUT FLOWERS

Walnut flowers

What do the walnut flowers look like?

The walnut blooms in May. Royal trees are a manioc unisex species, and within the same tree we encounter male and female flowers. The male flowers look bunched, long, thin and green; the feminine flowers are small, green, ovoid or spherical, covered by white, soft and dense brushes, which are present at the top two lobes and range from light green to bright, light red.

Male flowers, called imperious and buds, are grouped in inflorescences located at the basal part of the branches of the previous year and are composed of 100–160 flowers. The male flowers carry the pollen grains in the anthem.

On the other hand, female flowers are solitary, grouped in 2–3 flowers, rarely more, and located at the top of the shoots. Each flower has a hairless or florescent ovary at the extremity, where there is a large stigma consisting of two lobes. On the surface of the two lobes, there are numerous alveoli.

On the lobes, outside and inside, the humidity is very high, and the necessary conditions for the germination of pollen are ensured.

The walnut tree belongs to the group of anemophily species, those species whose pollen is transferred by wind within the same tree. Sometimes the pollen is lost when the male flowers release pollen into the wind, and the female flowers may be less or not at all receptive to the pollen.

Thus, although all varieties of walnut are asexual, the pole from the male flowers of a tree can travel to the female parts of the same tree and thus produce fruit without having other trees around. It is recommended that next to a walnut tree, there should be another variety of walnut tree, an early variety next to a later variety. This is desirable for obtaining higher nut production.

Using Walnut Flowers

Walnut flowers can be used successfully in dendrotherapy (healing with the help of trees). Obviously, it is desirable that walnut flowers are not broken, destroyed, or collected in large quantities to the detriment of fruit. However, there are cases where flowers are collected and used for their properties.

Hair Treatment

Who doesn't want to have clean, healthy, and shiny hair? A treatment that can help you heal your hair and scalp is the infusion of walnut flowers. All you have to do is pick 100 g of male walnut flowers and infuse them in water.

It is recommended that, in order to achieve the desired result, if the problems with the hair and scalp are considerable, then an infusion should be used daily for a week. Then, it can be reduced to two or three days a week, and finally, once a week. The duration of use depends only on each result.

Among the properties of walnut flowers is the support they give to the heart. Male flowers can be used successfully in the treatment of cardiovascular diseases as well as pancreatic insufficiency, as they lower cholesterol levels and normalise blood sugar levels.

Extract of Walnut Buds (Male Flowers)

The geotropic extract from male walnut flowers has various uses. It helps to reduce inflammation in the mucous membranes and skin (acne, seborrhoea), improve the functioning of the pancreas, help lower blood sugar levels and regulate bacterial flora, reducing bloating and flatulence.

The mode of administration varies and is dependent on the particular condition. The duration of administration can vary from one to three months, with intermittent breaks of one to two months. According to studies, no adverse reactions or contraindications are known.[7]

---

7   Information from Plantextrakt laboratories.

STIMULATING APPETITE

The following recipe can be used as a remedy for a lack of appetite and to gain weight. It can also be used to combat weakness, fatigue, nervous diseases and glandular and sexual insufficiencies. Iodine and vitamin B complex, which are present in the walnut "bud" and help the functioning of the thyroid, are also used in the prevention and treatment of endemic gout.

If consumed fresh, the preparation increases the body's resistance to infections. In small amounts (1 teaspoon per day), this preparation can also be given to children. In the case of adults, it is recommended to take a spoon before each main meal (3 times a day) for a month.

The preparation is obtained from:

Ingredients: 200 ml water
100 g of walnut "mugs"   50 g flowers

1. Flowers need to be washed first and then boiled over a low heat until the water drops decrease by half.

2. Boil with the lid on.

3. After boiling the flowers will wilt, leave them to cool and then add 50 g honey.

4. After mixing very well until fully homogenised, the contents should be placed in a coloured glass or ceramic dish. This will avoid direct contact with light. Hermetically sealed, the dish should be kept cold.

> **Walnut flowers can be used alongside other plants in prevention, improvement and different treatments.**

Although not smelly, colourful or aesthetically appealing, walnut flowers are each a slice of health offered by nature. From this arises the most complete fruit offered by nature, the walnut.

Chapter 5

ABOUT WALNUT FRUITS

MORE THAN 100 varieties of walnuts, with fruits growing in bunches (a bunch can have up to 15–20 fruits, each weighing up to 25g) have been discovered over time in Russia. The fruits of the walnut are the ones that give life, but they are also the ones from which life is reborn.

A drupe, a fruit with a fleshy mesocarp (sometimes juicy) and an endocarp (internal tissue that coats and protects the seed), consisting of a single kernel, is the type of fruit found in the walnut. Cherry, plum and peach are also drupes, but at a smaller size.

Until maturity, the seed is covered by a thin, soft skin, which must be peeled before the core is used. Consumed with the core, the skin leaves are bitter and unpleasant and have an iodine taste. Once mature, the seeds decrease in volume and dry out, and the skin sticks to the core. It can be eaten as it is.

Peel and Green Walnut

Nothing is wasted; everything is used. That is largely what is said about the walnut fruit. Whether we are talking about its use for homeopathic purposes, in food, cosmetics, or the textile industry, the development of the walnut fruit is, undoubtedly, the support that Divinity has to offer humanity.

People have learned over time to extract all the good properties from the green-covered walnut and use them for health purposes. This coating is very rich in iodine and tannins, having anti-infectious, purifying and stimulant effects on thyroid activity. It is no wonder, then, that a teaspoon of green walnut taken in the morning can work miracles.

SWEET GREEN NUTS

The youngest nuts will be chosen. They will be picked one by one from the tree. For the preparation of sweetness, it is necessary to have:

Ingredients:
1 kg peeled walnuts
1 kg white sugar
4 large lemons
200 ml water

1. Clean the green walnut peel with a sharp knife and place the nuts into a cold bowl of water until you finish peeling all of them.

2. Drain the cold water and pour the boiling water and leave the walnuts in for 10 minutes.

3. Wash with cold water and keep each time in cold water for 10 minutes.

4. Repeat this operation 4–5 times, taking care that the changed water is boiled each time.

5. In a separate bowl, add all of the sugar with 200 ml water and boil quickly over a high heat to avoid caramelisation.

6. When the syrup starts to thicken, drain the water from the walnut mixture and place the well-drained walnuts into the sugar syrup.

7. Once the walnuts and sugar syrup have mixed, add half a sliced lemon.

8. Allow the sugar to boil until it thickens, stirring often. The dish must be kept under constant watch as it can burn very easily at the bottom, and also the sugar can quickly give in to the fire.

9. To check if it is ready, take off the heat and place a few drops of syrup on to a plate. Once the syrup is cool to touch, if the mixture is firm and its viscosity is thick, it should not be put back on the heat. Pour into sterilised jars and seal.

Top tip: *If you add vanilla powder or liquid, during step 8, then the juice from half a lemon is added to the last boil, in step 9. This will add more flavour.*

Walnut shells are also useful, but there should be consideration given to how these are collected. Depending on the purpose for which green walnut strips are used, select only the pieces of a healthy green (those attacked by diseases or those yellow and past should be discarded), which are left to dry in a single layer with the hollow upwards, achieving optimal levels of water evaporation. When perfectly dried and brittle, they should be collected and stored in cloth bags.

PICKLED GREEN NUTS

This product is healthy, and despite the name, it cannot be consumed as a single dish or as an appetiser. It is a spice that is meant to improve your dish. For the preparation of the spice, you need:

Ingredients:
50 green nuts
150 ml brine (water and salt concentrates)
5 g coriander beans
5 g mustard beans
2–3 Peppercorns
3 red chilies

5 to 7 Cherry leaves (optional)
2–3 cloves garlic
100 ml Vinegar

Tools:
You will also need a needle and a hermetically sealed bowl or pot

1. Wash the nuts thoroughly and prick in random depths with a needle in several places.

2. Put them in the hermetically sealed bowl. This will prevent the green walnuts from changing their colour.

3. Separately prepare the brine.

4. Pour the brine over the nuts, stirring the contents from time to time.

5. After three weeks, remove the nuts and rise them passing the nuts under cold water and leave to dry for 2-3 days.

6. The combination of brine and air will give the nuts an almost black colour. After this period, the walnuts should be added to the bowl and the remaining spices should be added this time.

7. Finally, the dish should be filled with vinegar and the walnuts should be left to soften for at least two weeks. They can then be used for spicing your dishes.

Harvested green walnuts

WINE FROM GREEN NUTS

Green nut wine will not support you at a party, but it can help you afterwards! Although it is named after the product obtained from grapes, wine from green nuts cannot be consumed as an alcoholic product. Its purpose is therapeutic. One teaspoon taken three times a day, before the main meals, helps to detoxify the body and combat diarrhoea and other general weaknesses of the body.

To get a good cure, you will need:

Ingredients:

1 litre of distilled alcohol (over 40% strength)

2 litres of red wine

2 kg sugar

70 green nuts

Tools:

You will need a dark hermetically sealed bottle or container and a wooden spoon

1. Place the walnuts in a dark container, which can be hermetically sealed.

2. Then, pour the alcohol over the walnuts.

3. Leave to macerate for seven days.

4. At the end, add the wine and sugar together. Mix well until dissolved. It is preferable to use a wooden spoon to prevent oxidation.

5. Hermetically seal the dish and leave to macerate in a cool place, away from the light, for 40 days.

6. After this period, the product should be filtered, and the liquid obtained can be stored in dark bottles in dry places.

HYDRO ALCOHOLIC EXTRACT (TINCTURE) OF GREEN WALNUT SHELLS

Of all the preparations obtained from walnut, hydroalcoholic extract is the most active, being very rich in iodine, flavonoids and tannins. It is successfully used in the body's detoxification process (to remove oxidants, giardia, intestinal parasites, etc.), which contributes to the proper functioning of the thyroid gland (either in the case of hypothyroidism or hyperthyroidism). Also, it can be used in cases of hypothermia or urinary incontinence.

The contents should be consumed only after they have been diluted with water. Up to four teaspoons per day can be given to help relieve symptoms of rheumatism or gout.

Externally, the tincture can be used for disinfecting wounds.

The product is easy to prepare. It requires:

Ingredients:

300 g green peel powder   500 ml 50% food alcohol

1. Place 300 g of green walnut peel powder and 500 ml of food alcohol in a jar together.

2. After the hermetic closure of the jar, leave the contents to macerate.

3. The container must be stored in a warm place for a minimum of seven days.

4. At the end, the contents should be filtered, and the tincture obtained stored in dark bottles. It is preferable that the container in which it is stored is small in volume, so once opened the bottle should be able to be used more quickly and will not allow the product to come into contact with the air too many times, as this, can lead to its deterioration.

> **Walnut peels are known to have a powerful purifying effect, favouring the mobilisation and elimination of toxins from the body.**

NUT SYRUP

Some names given to preparations obtained from walnut shells (and beyond) may be slightly misleading as a mode of use. This is also the case with syrup from green walnut shells.

Syrup, by definition, is a dense solution, used as a soft drink (mixed with soda or mineral water) or as an addition when preparing a sweet dish. This time the purpose is to treat stomach ailments. You can take three teaspoons a day after the main meals.

But it is also known as an energiser, so be careful not to consume it too late in the day.

Here is what you need to prepare this syrup:

Ingredients:

150 g green walnut shells   2–3 cloves
1 litre water   1 kg white sugar

1. Boil 1 litre water with cloves.

2. Cut the walnut shells into pieces and immerse in the boiling water.

3. Allow 10 minutes to macerate.

4. When soft, strain the water into to a container.

5. Stir and add slowly 1 kg white sugar.

6. The obtained syrup should be put in dark bottles and simmered for another 10 minutes in a bain-marie or similar device, then it should be hermetically sealed.

COMBINED DECOCTION OF GREEN WALNUT SHELLS

This preparation is perhaps most often used in traditional medicine, given its versatility. It can be used for both internal and external purposes. It can be consumed in the case of intestinal worms (perhaps the best natural treatment in these situations), infections in the throat and mouth sores.

The preparation is a powerful antibiotic; it can be administered for remedies for up to 12 days consecutively by consuming from a half to one litre of the liquid per day. It can also be used in baths for cases of pain, bleeding, or the appearance of pinworms.

The ingredients of the product are simple:

Ingredients:	Tools:
4 walnut shells	You will need a dark jar
500 ml water	

1. Cut up 4 teaspoons of walnut shells and leave in a bowl with 250 ml of water at room temperature for 8 hours.

2. After 8 hours, the resulting extract should be filtered through a sieve and left in a pot.

3. The remaining walnut shells, after filtration, are boiled over a low heat for 5 minutes with the remaining water (250ml).

4. Touch a few walnuts randomly. If they are soft, they are ready.

5. The resulting product should be strained into the pot mentioned in step 3 and the two extracts will combine.

6. The final product should be poured into a dark jar.

MACERATE GREEN NUTSHELL

This recipe is recommended for external use only. It is beneficial in the treatment of varicose veins, dermatitis, or diseases of the musculoskeletal system.

It requires:

Ingredients:
50 g green walnut shells
200 ml apple vinegar

Tools:
A container that can hold at least 800 ml of liquid

1. Cut and chop the walnut shells and place in an empty container which approximately holds 800 ml.

2. Add the apple vinegar until the dish is filled and leave to macerate for 2–3 weeks, after which it can be used.

GREEN WALNUT PEELS WITH HONEY

Walnut peels, along with other plants, can increase their area of use. The following remedy is ideal for healing coughs if one teaspoon is given three times a day for seven days. In this preparation, green walnut shells are used alongside elderberries.

Ingredients:
4 whole walnuts
15 g dried elderberry
500 ml water

15 g honey

Tools:
You will need a grater

1. Boil 500 ml of water.

2. Grate the walnuts on the tiniest facing.

3. Mix with 1 teaspoon of dried elderflowers.

4. The mixture should be boiled in 500 ml of water over low heat for 15 minutes. If your cooker is an electric cooker, select which hob/facing you want to use and set the dial to 3.

5. Once cool, it should be filtered, and the extract should be mixed with a teaspoon of honey.

FRESH GREEN WALNUT PEEL

The green walnut peel can be used, without preparation, to combat warts. Gently rub the effective area with a fresh walnut shell. Do this for five minutes at least two or three times a day. Repeat the operation daily for one month; after that, results begin to appear.

> The thick green coating is not only used for treatments but, the green walnut peel is successfully used in the textile industry to produce natural black pigments (very resistant), in the cosmetic industry (in shampoo preparations), or the chemical industry (obtaining paints).

In green walnut shells, the vitamin C content is forty times higher than in orange juice and ten times more concentrated than in lemon juice. Remember that the shells are very rich in iodine and have disinfectant, purifying and mobilising properties to help aesthetically heal any lesions quickly.

There are also contraindications to the use of preparations from green walnut shells. It is not indicated for pregnancy, lactation, hyperthyroidism, irritable bowel and acid gastritis.

Like any natural treatment, any product should be administered only after a good understanding of the side effects and used after a discussion with the medical specialist, the homeopath or after carefully reading the leaflet, in the case of preparations bought from specialty stores.

Walnut Kernel

According to Dr Frank Hu, professor of nutrition and epidemiology at the Harvard School of Public Health, "We found that people who ate nuts every day lived a longer and healthier life than people who didn't eat nuts."

Coincidence or not, the walnut core is strikingly similar to a human brain: it is a miniature skull. The woody shell (cranial box) protects the nut core (brain), which has striations and circuses. It is no wonder, with such a similarity, that the walnut core has so many beneficial properties for the human body.

The nut's core is the most complex and rich fruit in fatty acids, vitamins and other beneficial substances. In the following, you will convince yourself why we need to have in our diet a substantial consumption of walnuts.

When fresh, the nut's core contains 17.57% water, 11.05% nitrogenous matter, 41.58% fat, 26.5% extractive materials and semi-wooden materials, 1.3% cellulose (organic substance of which the cell walls of plants are composed), 1.6% ash.[8] The energy value of the walnut core is, on average, 700 kcal per 100g.

Carbohydrate Content

Carbohydrates are natural organic substances that form the foundation of living matter and play an important role in metabolism. They perform a structural and energetic role from a biochemical and physiological point of view. Carbohydrate content in nuts varies between 11% and 14%.[9]

Protein Content

Proteins are complex macromolecular organic substances made up of amino acids, which are found in animal and plant cells. They are indispensable for human nutrition. In nuts, the content varies between 14–15%.[10]

8   V. Cociu, 2006.
9   D. Beceanu, 2004.
10 D. Beceanu, 2004.

Content in Protides

Protides are substances of great physiological and structural importance. Depending on the species, variety and tissue analysed, the content of fruit and vegetables may vary. Protides are found in walnuts in amounts ranging from 15 to 16.4%.

Content in Amino Acids

The building blocks of proteins are amino acids. Most of the amino acids we consume are directed to the muscles, and their main role is to ensure the toning of muscle mass. The walnuts are a powerful antioxidant; they have essential amino acids, which help to improve immunity and maintain a strong immune system. These qualities are only part of the role those amino acids have in our body. All the essential amino acids required are found in nuts.

Isoleucine – necessary, along with other substances, in the supply of body energy and brain stimulation.

Leucine – an indispensable amino acid. In the body, leucine enters the composition of proteins and intervenes in numerous chemical reactions.

Valine – ensures positive mental health, emotional control and muscle coordination.

Methionine – an amino acid essential for vital processes and used as medicine in liver diseases. Methionine is a source of sulphur, which prevents diseases of the hair, nails and skin; lowers cholesterol; protects the kidneys; is a natural chelating agent (i.e., it removes heavy metals that are toxic); and regulates ammonia levels.

Phenylalanine – allows the brain to produce norepinephrine, used in the transmission of electrical impulses between nerve cells and the brain; controls the feeling of hunger; is an antidepressant; improves memory and attention.

Threonine – decreases the fat production of the liver; stimulates the optimal functioning of the digestive system and metabolism. Threonine is also an essential amino acid that cannot be synthesised by the animal organism and can function as an alcohol.

Tryptophan – is relaxing, combats insomnia, prevents migraines, is anti-anxiety and anti-depressant; is an immune stimulant; reduces the risk of cardiovascular accidents; and, together with lysine, lowers cholesterol levels. An adult's daily requirement is about 0.25g.

Lysine – is an amino acid indispensable to growth. It has the property of dissolving red blood cells, bacteria and tissue cells, ensuring adequate absorption of calcium and participating in the formation of collagen (a component of bones, cartilage and connective tissue), antibodies, hormones and enzymes. A lysine deficiency causes fatigue, an inability to concentrate, irritability, slowed growth, hair loss, anaemia and reproductive problems.

Histidine – indispensable for the growth of mammals and is used in the treatment of rheumatoid arthritis, allergic diseases, ulcers and anaemia. A histidine deficiency can cause hearing problems.

Non-essential amino acids are also found in the walnut core and are important for the human body.

Arginine – an amino acid that exists in proteins and participates in the synthesis of urea in the liver, the source of nitric oxide. It increases the level of growth hormone, protects the walls of blood vessels, and reduces blood pressure. The usual dose is 5–10 g per day for adults.

Cystine – a non-essential amino acid that contains sulphur, is converted into glucose, produces taurine, and opposes excessive insulin production.

Tyrosine – a non-essential amino acid energising, anticatabolic, and helps the body build proteins. Tyrosine can support the production of thyroid hormones that will speed up your metabolism.

Aspartic acid – a non-essential amino acid, increases fatigue resistance because it works against ammonia, one of the toxic elements that our bodies produce.

Glutamic acid – used in medicine as a tonic for the central nervous system.

Glycine – also called glycerol, is a non-essential amino acid that stabilises the central nervous system, maintains prostate functions, and is converted into creatine to support the muscle cells.

Proline – ensures the proper functioning of joints and tendons and tones the heart muscle.

> In the eighteenth century, it was believed that consuming walnuts could cure headaches.

Lipid Content

Lipids are natural organic substances present in all living organisms. Also called fats, they produce heat in the body by oxidation. Walnuts contain on average about 62–65% lipids.

In walnuts, we find a high content of palmitic acid, 4.5%. By comparison, hazelnuts and almonds have a content of 3.0–3.3%. Palmitic acid is used in the manufacturing of soap, metallic palms, lubricating oils, waterproofing materials and food additives.

The nut fruits contain 1% of their weight in stearic acid. This is used in cosmetics in the preparation of vegetable butters, increasing the viscosity and stability of preparations such as emulsions, solid balms or shampoos, and to increase the firmness of candles or soaps.

Unsaturated fatty acids (found in the liver core) that cannot be synthesised by the human body are linoleic acid and arachidonic acid. For the human body, essential fatty acids are needed in amounts of about 7 g/day, much higher than vitamins. In practice, they are called vitamin F.

Vitamin Content

Vitamins are necessary to maintain vital cellular processes. These organic substances are found in different concentrations in the walnut core.

Vitamin A (retinols or axerophthol) – increases resistance to infections of cartilage, mucous membranes and skin; helps to adapt visually; helps the normal development of the skeleton and tooth enamel; as well as supporting the proper functioning of the liver, thyroid and other organs. It contributes to the regulation of sleep and blood pressure, intervenes in the metabolism of proteins and mineral salts, especially calcium, fats and carbohydrates, and plays an important role in the care of the skin and mucous membranes, as well as in stimulating growth. The daily requirement of vitamin A is 1,800 (1,100–2,500) I.U. in children and adolescents; 5,500 (5,000–6,000) I.U. in mature women; 6,000 I.U. in pregnant women; and 8,000 I.U. in lactating women. When the body is weakened, as during the period of diseases that are accompanied by fever, the need for vitamin A increases. The walnut kernel has a 4% amount of vitamin A.

Vitamin B1 (aneurin, orizanin, tianin) – has an important role in nervous balance, stimulating the activity of the central nervous system, helping to burn carbohydrates and proteins, normalising the function of the nervous system, and participating in the growth process. Avitaminosis B1 can cause beriberi disease with paralysis of the lower limbs, which can stretch to the back muscles, heart disorders, rheumatic pain, cranial nerve disorders, vomiting, seizures, syncope and death by asphyxia. Vitamin B1 helps the nervous system and heart function ("heart vitamin"), stimulates growth, and aids digestion. Vitamin B1 overdoses affect the thyroid gland and insulin secretion, generating vitamin B6 deficiency and loss of other B vitamins.

Vitamin B2 (lactovina or riboflavin) – stimulates growth, participates alongside vitamin A in the process of vision. For healthy adults, the daily requirement of vitamin B2 is 1.21.8 mg. In general, a quantity of 0.6–0.7 mg vitamin B2 is recommended for every 1,000 calories, during periods of pregnancy and breastfeeding, women need 1.5–3 mg of vitamin B2 per day. In children, depending on age, the daily requirement of vitamin B2 is 0.4–1.2 mg. Alcoholism, diabetes and various antibiotics can reduce the intake of this vitamin. The existence of Hypovitaminosis B2 is not known. In the walnut core, there is a quantity of 0.12 mg% vitamin B2.[11]

11   W.S. Souci et al., 1981, C-tin Pârvu, 2006.

Vitamin B3 (nicotine-mida; vitamin PP; niacin) – is indicated in the fight against pellagra (a disease that occurs as a result of long consumption of corn, excess alcohol consumption, prolonged exposure to the sun, as well as the intensity of physical exertion), herpes, acne, pruritus, frostbite, sores, oral inflammation, enterocolitis, various psychoses, depressive states, arterial hypertension, atherosclerosis, difficult breathing, alcoholism and stress. The deficiency is manifested by apathy, weight loss, drowsiness and decreased reflexes. All these symptoms result in the appearance of pellagra, which can become a digestive disorder that could induce pain or diarrhoea. A deficiency of vitamin B3 can cause delirium, disorganisation, dementia, or neuropsychic disorders manifested by melancholy. A lack of vitamin B3 affects the skin, making it more sensitive to sunlight. As a result, the skin turns red, and then blisters, cracks, dermatitis, etc., can appear. The content of vitamin B3 in nuts is 1.00 mg%.

Vitamin B5 – produces energy, invigorates the body and reduces stress. Vitamin B5 deficiencies are more common in those who consume alcohol often, those who have high cholesterol, and the elderly. The daily requirement of vitamin B5 is 5–15 mg. A normal diet ensures a daily intake of 5–20 mg pantothenic acid, which in some proportion can also be synthesisd by the intestinal flora. Walnut kernels have a content of 0.82 mg% of vitamin B5.

Vitamin B6 (pyridoxine; adermin) – has a beneficial influence on the skin as well as in the growth processes. Hypovitaminosis B6 occurs infrequently and leads to various nervous disorders (irritability, mental liability, general weakness, walking weight, etc.), as well as the appearance of skin diseases (premature curling of the skin, hair loss, tingling and numbing of the extremities, abdominal pain, etc.). The body's deficiency in vitamin B6 occurs in diets with excess protein, in absorption disorders, in chronic alcohol poisoning, high blood pressure and myocardial infarction when oral antibiotics are abused. For adults, the daily requirement of vitamin B6 is 1.6–2.5 (3) mg, directly proportional to protein intake. For women, during pregnancy, the daily requirement of vitamin B6 is 4–6 (7) mg, but it is also 4–7 mg for women who have leukaemia or who take birth control pills. The maximum losses in vitamin B6 by culinary processing are about 40%. The content of vitamin B6 in the walnut core is 0.08 mg%.[12]

Vitamin B8 (vitamin H; biotin; coenzyme R) – relieves liver functions (helping to remove toxins from the body) and gallbladder; facilitates the proper use of fats and cholesterol by the body; prevents the accumulation of fat in the liver. Recommended daily: 0.3 mg (300 mcg). Children need 50–90 microgram mcg/day. In the walnut core, there is a 0.02 mg% quantity of vitamin B8.[13]

Vitamin B9 (plococ acid; pteroylglutamic acid) – is indispensable in the formation of red blood cells, preventing anaemia. Promotes emotional balance, decreases pain, increases the production of breast milk, and prevents the appearance of cancer in the lungs, the colon and the uterus. Protects the body from cardiovascular disease and strokes. A vitamin B9 supplement is indicated for people who consume excess alcohol.[14] The recommended dose is 0.4 mg per day, or 100–200 gamma (1000th of a milligram) in 24 hours. The nut content of vitamin B9 is 0.08 mg%.

---

12   C-tin. Pârvu 2006.
13   C-tin. Pârvu 2006.
14   C-tin. Pârvu 2006.

Vitamin C (ascorbic acid) – the most powerful antioxidant. Vitamin C has anti-infective, anti-toxic and toning properties. Ascorbic acid also participates in the assimilation of iron by the body and prevents and heals scurvy. It also increases the resistance of blood vessels, contributes to the formation of red blood cells, teeth and bones, has the role of regulating blood sugar and cholesterol levels, destroys toxins accumulated in the body, etc. The first symptoms of a deficiency of vitamin C could be: frequent colds, a tendency toward inflamed mucous membranes, varicose veins, haemorrhoids, excess body weight, persistent fatigue, nervousness, difficulty concentrating, depressive states, sleep disorders, wrinkles, skin wrinkles, hair loss, general weakness and a general lack of charm and charisma. A healthy adult's requirement for vitamin C is 73 mg/day. For women who are pregnant or during pregnancy, it is recommended that the daily dose of vitamin C be 100 mg/day, and in lactating women, it is 50 mg/day. The dose for an infant is about 30 mg of vitamin C every 24 hours. The need for vitamin C in children is 1.5–2 mg/kilogram-body/day, depending on age. In children between 1 and 14 years, the recommended dosage is 30 to 90 mg every 24 hours; between 15 and 20 years, the requirement is about 100 mg over a 24-hour period. Vitamin C is not toxic, but it is destroyed by boiling fruits and vegetables. It is better to consume fresh vegetables and fruits to absorb as much vitamin C as possible (remember to wash the vegetables and fruits all the time before you eat them or prepare them). The nut content in vitamin C is 3 mg%.[15]

Vitamin E – intervenes favourably in reproduction, ensuring the normal functioning of the sex and endocrine glands, and it aids in the storage of glycogen in the liver and muscles, including in the heart muscle. Vitamin deficiency produces sterility in men and it decreases sexual instinct. It can cause

15   15 mg%, after C. Pârvu, 2006.

testicular, ovarian and uterine atrophy. In children, it causes delayed development of the genital organs and delayed puberty; menstruation is accompanied by sometimes violent pain. It is a complex of vitamins necessary for men, with the daily dose being 10–25 mg. For athletes, an administered 50–70 mg daily dosage can increase performance to 90–120 mg/day, and during training and at the big competitions, 150–200 mg/day. The content of the walnut core in vitamin E is 24.70 mg%.

> The walnut kernel and its curative properties have been known and used since the time of the Greeks and Romans. There are writings that show the attributes and properties of walnuts (see Pliny the Elder). The list of remedies for which nuts were used in antiquity is long and starts with the treatment of inflammation, scarring, burns, animal bites and so on.

As a nutrient, the walnut core has a calorie content of 55.8, and the energy value is 654 kcal. The walnut core remains the fruit, with a calorie content of lipids of 545.8 kcal, and the protein has an intake of 52.8 kcal. In the walnut core, there is 0.06 g of starch.

The walnut, however, is the only fruit that has in its structure almost all the chemical elements necessary for the body.

Phosphorus – controls calcium balance as well as acid-base balance; comes into numerous combinations with proteins, lipids and carbohydrates; acts favourably with vitamin D and calcium; takes part in the formation of bones, teeth and blood; has a special importance in the production of nerves, intellectual and sexual energy; regulates the heartbeat; and helps the normal functioning of the kidneys. In the walnut core, there is a quantity of 346 mg of phosphorus, with the recommended daily dose being 35%.

Potassium – helps to preserve the acid-base balance of cells, revitalising the body and thus maintaining its health; increasing diuresis and sodium elimination; promoting neuro-muscular excitability; also intervening in the processes of permeability and in the metabolism of carbon-hydrates; being a cardiac tonic and a stimulant of intestinal movements; contributing to the proper functioning of the adrenals; the elimination of organic residues; and eliminating excess potassium through the kidneys. Potassium deficiency produces hypoglycaemia, oedema and degeneration of the heart muscle and skeletal muscles, negatively influencing growth. Among the enemies of potassium are alcohol, coffee, sugar, diuretics, physical and mental stress and, finally, smoking. Walnut kernel has a content of 441 mg of potassium, and the recommended daily dose for this element is 13%.

Calcium – is an essential element for the human body, especially in the first years of life when the process of forming bone tissues takes place. Calcium is indicated in cases of fatigue, in various lung diseases, in adenitis, asthenia, headache, irritability, nervousness, rickets, thinning of bone structure (osteomalacia), acute or chronic spasmophilia, osteoporosis, various allergic reactions, menopausal disorders, prolonged bleeding, etc. Calcium deficiency produces tooth decay, muscle tetany, insomnia, osteomalacia and a tendency to break bones (especially in women with multiple births). Calcium is found in the walnut core in a quantity of 98 mg, and a daily dose of 10% is required for the body.

Magnesium – plays a role in the formation of teeth and bones; along with calcium and phosphorus, takes part in excitability control; also plays an important plastic role, representing one of the main constituents of the skeleton and teeth; regulates calcium balance; helps the body's muscles function properly; strengthens the body's defence reactions to infections; delays ageing and alleviates the pain caused by senility; activates the neuromuscular system; regulates the body's temperature. Magnesium deficiency manifests itself through spasmophilia, hyper-emotionality, a feeling of a lack of air, headache, tremors, dizziness, insomnia, muscle cramps, tingling and numbness of the extremities, eye fatigue, palpitations, biliary dyskinesia, abdominal colic, brittleness of nails, hair and teeth, depressive and hysterical states, etc. 158 mg is the magnesium content of the walnut core, with a daily dose recommendation of 40%.

Sodium – also intervenes in maintaining an acid-based balance by stimulating muscle mass. It also prevents sunstroke and contributes to the proper functioning of the nervous system and muscles. A 2-mg quantity of sodium is found in the walnut core.

> It is important to keep in mind the beneficial effect of the walnut core will only be achieved when chewing the core for a long time, so that the body can receive all the nutrients.

The walnut core has several microelements in its composition (chemical elements that are found in small amounts in the body or in the soil), which enter the composition of hormones, enzymes, etc., and whose absence can lead to growth disorders, reproduction disorders and nervous disorders.

Iron – an important anti-anaemic factor, the main constituent of haemoglobin in the blood, some enzymes, etc. – is indispensable in the transport of oxygen in the body as well as in cellular respiration. Iron deficiency produces iron deficiency anaemia and stomatitis (inflammation of the oral cavity). Many infants suffer from anaemia due to insufficient iron in their food. On the other hand, excess iron contributes to the lowering of the level of copper in the body, which can cause poisoning. A dose of 2.8 mg or 2.91 mg of iron is found in the walnut core, and a daily dose of 16% is recommended.

Copper – is indispensable in cellular respiration and bone formation; it is also particularly important in that it helps fix iron, forms haemoglobin and red blood cells, is resistant to excessive blood clotting, etc. Copper poisoning can be manifested by jaundice, anaemia, the need for excessive water consumption, etc. In the walnut core, there is a quantity of 1.586 mg–1.6 mg of copper; a daily dose of 79% is recommended.

Manganese – regulates the growth and metabolism of proteins, lipids and carbohydrates; promotes liver and kidney functions; accelerates burning, helps fix calcium and iron; acts as a vitamin; plays a role in bone formation; activates a large number of enzymes; contributes to cholesterol synthesis; plays a decisive role in the activity of the endocrine and sexual glands, etc. Manganese is indicated in the regulation of blood pressure, in arthritis, non-infectious asthma, migraines, dysmenorrhoea, allergic eczema, hives, gout and rheumatism, thyroid disorders, morning asthenia, fatigue and nervousness, etc. Manganese deficiency is manifested by anaemia, sterility, and in some cases, decreased bone strength and the appearance of skeletal malformations, impaired reproductive function, convulsions, paralysis, changes in lipid and carbohydrate metabolism, noises in the ears and even deafness, difficulty coordinating movements, etc. The recommended daily dose of manganese is 1.7 mg, with the walnut core having a concentration of 3.4–3.414 mg.

Zinc – has an important role in maintaining visual acuity, protein metabolism, activity of the pancreas and sexual organs, hastening the healing of burns and wounds, stimulating the process of assimilation of vitamins, forming leukocytes, activating the immune system, favouring the synthesis of nucleic acids and the regeneration processes, etc. Zinc also stimulates mental activity and helps the brain function properly. Zinc deficiency affects sexual growth and development, decreases appetite, accelerates the rate of ageing, leads to anaemia, hair loss and nail deformity, liver disease, repeated infections, decreased taste and olfactory sensitivity, difficulty in wound healing, various neuralgias, the emergence of prostatitis, damage to the immune system, sexual impotence, decreased sperm production and disorders of the reproductive system, etc. In the walnut core, there is a content of 3.09–3.1 mg of zinc, and the daily recommended dose is 2.1 mg.

Selenium – has antioxidant properties, which play a role in stimulating immunity, helping male fertility, reducing the risk of pregnancy and contributing to the synthesis of the thyroid hormone. Selenium, along with vitamins C and E, helps strengthen immunity and provides protection against viral infections. Some studies show that selenium is helpful in controlling the herpes virus, while others claim that it reduces the risk of Acquired Immune Deficiency Syndrome (AIDS) after HIV infection as it reduces the replication force of the virus in the body. Moreover, people infected with HIV who have selenium deficiency have a 20% higher risk of getting AIDS compared to infected people who have an optimal level of selenium in the body. For this reason, selenium is considered necessary as an adjunct in HIV/

AIDS therapy. In men, about half the amount of selenium in the body is concentrated in the testicles, seminal canals and prostrate area. Selenium is vital for sperm mobility, which is why it increases the chances of conception and has a positive effect on male fertility. Selenium deficiency is associated with cardiovascular disease and other conditions that cause inflammation in the body. In the nut's core, there is a quantity of 4.9 mcg, and a dose of 7 mcg is recommended.

Iodine – indicated mainly in the proper functioning of the thyroid gland, but also in conditions such as rheumatism, premature ageing, thyroid infections, hypercholesterolaemia and atherosclerosis, arterial hypertension, obesity, circulatory disorders, tuberculosis and other lung disorders, rickets, lymphatism, bronchitis, scrofulous, endemic gout and various dermatitis, hair damage and nail problems. Being one of the constituent elements of the thyroid gland, its insufficiency in eating causes cretinism and endemic gout. Particularly sensitive to iodine deficiencies are school-age children. Women who have had iodine deficiencies during pregnancy expose the baby to a risk. Often, a new-born can suffer from different conditions because of the lack of iodine received from the mother during the pregnancy. For adults, the daily iodine requirement is 100–130 mg; according to some authors, this need can reach 300 mcg/day or even more. During the period of breastfeeding, the daily requirement of iodine is estimated at 125–150 mcg. It should be noted that raw cabbage consumed in excess can cause an iodine deficiency, thus affecting the production of hormones by the thyroid gland. Excess iodine is not found in regular diets. About 30 g of iodine are found in 100 g of walnut kernels.

It should be noted that the walnut kernel is a complete and concentrated food, a real energy bomb. It has a content of fatty substances that varies between 52–70%, protein substances between 12–25%, sugars 5–25%, mineral substances 1.3–2.4%, etc.

Walnut kernels have significant amounts of antioxidants, Omega 6 fatty acids, and Omega 3. Omega 3 and Omega 6 help blood vessels and arteries (in that they become softer), produce more fluid blood, lower triglyceride levels, and reduce the risk of a heart attack. According to studies, consumption of foods rich in Omega-3 at least once a week reduces the risk of Alzheimer's disease (doctor's advice).

> It has been calculated that 1 kg of nuts is equivalent to 1 kg of bread, 0.5 kg of meat, 0.5 kg of fish, 0.5 kg of potatoes, 0.5 kg of prunes or 1 kg of pears. (V. Cociu, 1983). The walnut is a complex food, very high in calories and very high in energy 636 calories present per 100 g of core, which is especially valuable for anaemic children, the elderly, convalescents, diabetics and neurotic people.

But like any food that has so many beneficial properties, there are limits. Thus, the recommendation of the specialists is that the walnut should be consumed within the limit of 100 g per day (some specialists even recommend the limitation of 30 g per day, depending on the walnut variety but also on the characteristics of each consumer). Although only 400 grams of nuts can satisfy the daily need for food, providing all the necessary elements and vitamins, the unlimited consumption of this concentrated fruit, over 500 grams per day can be harmful to the body. Inflammation of the tonsils, rashes in the mouth, muscle spasms, and headaches are effects that can occur in the case of excessive consumption of walnut kernels. Although researchers at Yale University have found in studies that walnut kernels can lower cholesterol levels in the body, a real health hazard can occur if we do not store nuts properly. They can be infested with aflatoxins, a mould that can even lead to the development of some forms of cancer.

The walnut kernel is very versatile, which makes it very easy to use in various fields.

Walnut kernel

Walnut Kernels used in the kitchen

Even if processed, walnut kernels retain much of their beneficial properties, such as vitamins, lipids, or proteins, for the body. Obviously, the values decrease depending on how the walnut kernel is prepared, but even so, it remains an important source of health for the body.

In our diets, walnut kernels are a perfect substitute for meat (although it has the same caloric power as meat, walnut kernels are a much healthier food medicine) and better than soy. The walnut kernel given through the mincer can perfectly replace the minced meat in almost any preparation, from meatballs to rolls.

WALNUT KERNELS

Ingredients:

100 g of garlic

50 g walnut kernels

5 g salt

Tools:

Grater

Pestle and mortar

1. Peel the bulbs of garlic and grind well with salt. Use the mortar and rub until smooth, and the garlic will turn white like a foam.

2. Grate the walnuts well to a powder.

3. Add the ground walnuts over the garlic.

4. Rub until the paste is smooth and increase in volume.

This can be served as a garnish for various dishes.

Ingredients:

200 g nuts   10 g salt (optional)

1. Place baking paper in a tray and add 200 g of walnuts.

2. Place in the oven for 7 minutes at 150°C.

3. You can add salt sprinkled over the nuts from the beginning.

Also, walnuts can be fried over low heat, in a frying pan, on the cooker.

ADVICE FOR A RICH BREAKFAST

Combine walnut kernels with 50 g of whole grains and any type of milk. Alternatively, you can prepare a quick breakfast with whole-meal toast lathered with honey and sprinkled with baked nuts and tea.

WALNUT KERNELS AND SWEET POTATOES

Ingredients:

2 potatoes (any type)

30 g nuts

2–3 g cinnamon

1. Boil the potatoes and then cut them into cubes /4-5 inch each.

2. Cut into large pieces the walnut kernel adds them over the chopped potatoes and mix.

3. Finally, the cinnamon should be added and can be consumed as such.

Advice: If cinnamon is replaced with basil or parsley, then this preparation can also be used as a side dish for any meat or fish dish.

CABBAGE WITH WALNUT KERNELS

Ingredients:

200 ml double cream

100 ml yogurt

50 g ground walnut kernels

100 g grated cheese

1 small link of green parsley

12 g salt

4 g Pepper

4 g curry powder

4 g paprika

1 cabbage

1. Add the water to a pan and bring to a boil.

2. Remove the leaves from the cabbage, the yellow or damaged ones, and wash the cabbage.

3. In the pot with boiling water, add the remaining cabbage leaves and boil.

4. Once the leaves are soft remove them, drain well and cool them.

5. Cut the leaves into four equal parts.

6. Ground the walnut kernels very finely.

7. Separately, mix the cream, yogurt and cheese together.

8. Add the walnut kernels gradually until a creamy paste is obtained.

9. Add the spices over the creamy paste.

10. The cabbage pieces should be added to the cooking pot, and the walnut cream mixture should be poured.

11. Put everything in a non-stick tray and into the oven, which was heated to 180°C, for about 15 minutes.

12. After removing from the oven, while it is still hot, sprinkle with finely chopped parsley.

WALNUT KERNEL HUMUS

Ingredients:

200 g walnut kernels
150 ml walnut oil, cold pressed
A clove of garlic
100 g chickpeas (to be prepared in advance as below, or use precooked chickpeas)

2 g bicarbonate
2 orange peel
125 ml orange juice
g ground black pepper
g salt

1. Bake the walnut kernels in the oven for 8 minutes at 150°C.

2. Let them cool until the kernels are no longer hot and you can pick them up with your hands.

3. In a mixer, add the baked walnut kernels with 3 tablespoons of cold-pressed walnut oil and a clove of garlic.

4. Mix until the composition becomes creamy.

5. Add the chickpeas to the composition.

6. It is time to add half a teaspoon of grated orange peel, a quarter cup of freshly squeezed orange juice, salt and pepper (if desired).

7. The whole mixture should be mixed until it becomes creamy.

The result can be served with slices of bread or salads.

PREPARED CHICKPEAS

Leave the chickpeas to soak overnight in baking soda water, wash and boil the next day until the grain is crushed, drained and ready to use.

CREAM OF MUSHROOM AND WALNUT SOUP

Ingredients:

1 kg mushrooms
250 g walnut kernels
250 g onion
l olive oil

1 tsp salt
1 g thyme, chopped (fresh or dry)
1 g ground black pepper
125 ml soy milk

1. Chop the fresh mushrooms into thin slices.

2. Chop half an onion into fine slices or diced pieces.

3. Boil the olive oil and water (a tablespoon of oil and a tablespoon of water) for 5 minutes.

4. Add thyme to the pot.

5. Cover and simmer for another 5 minutes.

6. Add the walnut kernels and 125 ml of soy milk to a blender and mix until the mixture is creamy and the walnuts are well crushed.

7. Add everything into the mushroom soup mixture and simmer over low heat.

8. Leave to thicken for 5 minutes. (It is necessary to thicken it a little.)

9. Once it has thickened to your preference, you can add pepper and it is ready to serve.

VEGAN MEATBALLS WITH WALNUTS AND SEEDS

Ingredients:

400 g sunflower seeds ground

200 g seeds 4 mixes

750 g ground nuts

30 g oatmeal

4 finely chopped onions

40 g breadcrumbs

400 ml water

15 g basil

10 g thyme

5 g sage

3 g salt

1. Add water to a pan. When simmering, add diced onions and cook until they become transparent.

2. Once ready, add a mixture of sunflower seeds, nuts, breadcrumbs and oatmeal over the top.

3. Let them cook on the stovetop until the mixture of seeds and nuts is soft. Steer constantly.

4. Take the mixture from the hob. Mix, and then form balls or burgers.

5. The obtained buns should be placed on a tray lined with baking paper.

6. Place them in the oven heated to 200°C, and leave until they become hard and brown.

7. If the composition is too hard, you can add a little water. Its consistency must be soft enough so that it can be easily shaped.

These can be served as is, or with various sauces.

BEE HONEY WITH PRESERVED WALNUTS

Ingredients:

700 g bee honey

400 g walnut kernels

1. Add the walnut kernels to a jar of any type of bee honey.

2. Turn the jar upside down so that the honey penetrates everywhere.

3. It can be consumed as is, or in sweet dishes, and can be stored in the refrigerator or at room temperature.

Preserved walnuts and honey

NOODLES WITH WALNUT KERNELS

Ingredients:
500 g of raw noodles
200 g ground nuts
200 g of powdered sugar
5 g of vanilla sugar
Grated peel of 1 lemon
3 g salt

1. Boil the noodles in salted water until they are al dente.

2. Drain and wash with lukewarm water and set aside.

3. Create a mixture including ground walnuts, sugar, vanilla and lemon peel.

4. On a large plate, place a layer of noodles, one of walnuts, and finally sprinkle walnut kernels on top.

5. Put in the oven for 5 minutes at 200°C.

WALNUT KERNEL ROLL

VEGAN

Ingredients:

2 chicken eggs
200 g white sugar
180 g double cream
25 g ammonium
bicarbonate

700–800 g flour
200 g butter (margarine)
150 g walnut kernels
10 g vanilla
1 medium sized lemon

1. Separate the eggs.

2. Beat the yolks well with the sugar, vanilla, cream and baking soda until thick.

3. Grate the lemon and add the peel into the mix and stir.

4. Add the butter and flour, folding constantly.

5. Leave the resulting dough in the refrigerator for 30 minutes.

6. Grind the walnut kernels well and mix with the sugar.

7. Once the dough is ready, divide it into two parts and shape it into balls.

8. Roll the balls out the sheets, add the filling and roll.

9. Finally, grease each roll with beaten egg and sprinkle with walnut kernels.

10. Bake for 30–40 minutes at 180°C.

Walnut crackers

SANDWICH WITH WALNUT KERNELS AND CHEESE

Ingredients:

2 thick slices of any bread
5 Sliced cheeses (any cheese)
15 g walnuts
1/2 apple (red, white or yellow)
1 lettuce
1 fresh cucumber
1 fresh tomato
1 red onion (optional)
Butter
Black pepper, ground, to taste

1. Grease each slice of bread with butter.

2. Cut the walnuts into pieces, the apple into slices, the lettuce into thin strips, the tomato, the onion and the cucumber into slices.

3. Sprinkle the bread with walnut kernels and peppers; add cheese and apple slices; and top with lettuce, cucumber and tomato slices.

4. Add the other slice of bread on top, like a sandwich.

5. It can be fried for a few minutes on a hot tray or placed in the oven at 180°C for 2 minutes.

WALNUT KERNEL OIL

This oil can only be obtained industrially by cold pressing. The oil obtained is rich in fatty acids (over 50% linoleic acid).

Administered 3–4 times a day with a teaspoon, preferably on an empty stomach, it has a strong cholesterol-lowering effect. It is a very good vascular tonic and cleanses the bowel of tapeworm. It is also recommended to fight bronchitis, cornices, tuberculosis, constipation and syphilis.

Walnut oil can also be used in veterinary medicine. It can be given to cattle in cases of constipation or bloating.

When buying nuts, it is better to choose large fruits with a thin skin. The elongated oval fruits (the skin is rounded and thicker) have a smaller core. (It all depends on the variety.) The shell must not be cracked, damaged, or scratched. The good cores are dense and elastic, with a golden hue, and covered with a thin layer.

The walnut kernel is also used in the manufacture of technical oil, used in painting, and in the manufacture of luxury soaps, printing ink, etc.

Leaves and green walnuts

> I called the walnut a "pharmacy", and I do not think I am wrong at all. Its use for homeopathic purposes has been known since antiquity.

Naturalist's Advice

Take three pieces of walnut kernels along with 1 tablespoon of bee pollen and mix well in a circular motion. This mixture, taken daily, in the morning and on an empty stomach, prevents vascular accidents, strengthens blood vessels, and prevents them from breaking, especially in the elderly.

Daily consumption of 20–30 nuts accelerates healing of skin diseases such as psoriasis, allergic refractory eczema and fungal infections of the skin. It helps skin recover after burns and stops degenerative processes.

Walnuts, like hazelnuts, are very low in sugar, which makes them a staple in the diet of those with diabetes, especially where it is recommended to eat mostly raw vegetables or give up meat. As with hazelnuts, the walnut kernel has a hypoglycaemic action and thus contributes to the relief of the disease.

> Due to its high fat content, ability to assimilate easily, and fact that it does not increase cholesterol, the walnut kernel helps to increase body weight. If you want to gain weight without risking an increase in cholesterol, an overload of the cardiovascular system, or the appearance of problems such as stretch marks or cellulite, then you can confidently consume walnut kernels.

Walnut kernels keep our minds active, reduce mental stress, and help us fight depression. Due to their high magnesium content, walnuts have calming and slightly sedative properties. Two to three nuts eaten with honey will remove brain vessel spasms.

Ten crushed walnut kernels together with a teaspoon of cucumber seeds are recommended for women at risk of losing a pregnancy. This paste is recommended with a mixture of water. Again, however, medical advice should be sought before exploring homeopathic remedies.

In the case of children who consume walnut kernels, they generally experience harmonious growth, and the development of intellect and physical tone is stimulated.

Abdominal colic in children is relieved if they take a well-dried and grated walnut kernel, mixed with a little pepper.

> If you have trouble sleeping, then consume fewer walnuts than the recommended day dose before bedtime. Walnut kernels are rich in melatonin, a hormone that helps you get a good and restful sleep. Earl Mindell and Hester Mundis praised the properties of melatonin in the *New Vitamin Bible*.

GROUND WALNUT KERNELS, MIXED WITH FLOUR AND DOUBLE CREAM (OR FATTY YOGURT)

The recipe is used to treat infected wounds (boils, eczema, abscesses and phlegmons) by applying the mixture to areas with infected wounds 2–3 times a day for 3 consecutive days.

This preparation helps to collect pus. It also has a skin function in cases of burns, sunburn, or excessive itching.

In the case of rickets and anaemic children or those with bone disease, it is recommended to massage with walnut kernel oil.

WALNUT KERNEL PASTE

1. Take the core from 2–3 large walnuts, grind evenly.

2. Mix with a few teaspoons of still water until a white paste is formed.

This is recommended for invigorating hair and maintaining oily skin. Applied by a light massage with circular movements. This helps to refresh the skin, invigorate it with blood, and provide deep cleansing by removing the layer of dead cells. For the hair, massage the hair and scalp, and then leave the mixture to act for 10 minutes. Then, rinse well with water.

> A portion of only 28 g of walnut kernels provides 7 g of protein, which corresponds to 13% of the recommended daily dose, which is why walnuts are also known as "vegetarian meat". For comparison, a regular egg provides 5.5 g of protein.

Be very careful how you store the walnut kernel and for how long you use it. Once the nut becomes rancid, this causes the loss of its nutritional qualities, but it also influences health in a negative way. Rancid fats affect the blood vessels and, implicitly, the heart. Glycerine from oil seeds is very dangerous. It is converted to acrolein, a substance that is toxic to the liver and has a carcinogenic effect. Once in the body, acrolein behaves like a free radical and is very toxic. Rancid nuts are not necessarily mouldy.

They easily change colour, taste and smell. If you taste them and they do not taste pleasant, do not eat them!

> Dried walnut shells can be used as fuel. They generate heat that lasts over time. The smell of burning walnut shells is also pleasant and fragrant.

Chapter 6

WALNUT LEAVES

T HE OLDEST walnut in the world exists in China. On the plaque at its base, it is written that it was planted between 1066–1600 BC, during the Shang Dynasty. The tree has a height of 25.4 m and a trunk thickness of 15.7 m. Its crown provides shade over a radius of a few hundred square metres.

The leaves of the walnut are large, uneven, compound, with 5–9 elliptical leaflets and globous edges. Walnut leaves contain elegiac tannins, plus a variety of naphthoquinone derivatives, caffeic acid, ascorbic acid, flavonoid glycosides, p-coumaric acid and a small amount of volatile oil.

> It is preferable that the walnut leaves that you want to use for various remedies be picked when they are still green, by detaching them from the branches and not collecting them from the ground. They will be harvested without stalks and dried in a shady place, placed in thin layers. If you want to keep them for a long time, keep in mind that walnut leaves lose their properties after two years of collection. To avoid their deterioration, it is good to keep them in damp, clean places and in paper bags or boxes.

Walnut leaves contain small amounts of essential oil, about 3%, a p-coumaric similar to the levels of oak bark, as well as hexyl-cyclohexane, gallic acid, ellagic acid, inositol, alpha- and beta-hydrogen and vitamin C. Below this brief description of botany lies a real contribution that they make to humans through the healing properties of the walnut leaf.

WINE FROM WALNUT LEAVES

Strange as it may seem, walnut fruit wine is healthy and a good substitute for grape wine. This procedure can take 5-7 days.

Ingredients:
500 g walnut leaves
1 kg white sugar
500 g of any type of bee honey
1 teaspoon of acid mixture (citric, malic, tartaric acid)

4 g of yeast nutrients
3.5 – 4 L of water

Tools:
Strainer or sieve

Walnuts on leaves

1. Boil together the wine yeast and the walnut leaves.

2. Gradually add the sugar and honey.

3. Boil for 30–40 minutes.

4. After boiling, the contents should be strained.

5. Separately, mix the yeast with the acidic mixture.

6. Add the wine yeast on top of the mixture.

7. Finally, strain the contents and leave to ferment in a dry, airy space for 2 weeks. Check daily to see the process.

It is preferable to place the container in a warm location with adequate ventilation so that it can ferment. The wine obtained will then be transferred to bottles, hermetically sealed, and kept in a cool place for 6 months.

Bottling will be done after this period into other bottles, hermetically sealed, and stored for another 6 months until consumption.

WALNUT FRUIT OINTMENT

Ingredients:
15 g Walnut leaves

100 g sunflower oil (cold pressed)
15 g beeswax

1. Chop the walnut leaves into small parts.

2. Let them to soak in the oil for a week

3. After a week, simmer the preparation for 30 minutes.

4. Strain and add beeswax.

5. Stir everything to a boil in a water bath (bain-marie) for 30 minutes.

The obtained paste is used to heal wounds, sunburns and oily skin. The product should not be used internally; use only as a topical cream.

WALNUT LEAF INFUSION

Ingredients:

Walnut leaves   1 L of water

1. The leaves should be crushed until their volume is equivalent to 3–4 teaspoons.

2. The leaves should be left to soak with ½ litre of water for 8–10 hours, after which the liquid should be strained and the leaves retained for further use (see step 4).

3. The resulting liquid should be set aside in a dark bottle.

4. The leaves should be boiled in half a litre of water.

5. Let it cool and then strain, discarding the leaves.

6. Mix the two extracts obtained.

The preparation obtained can be used in the case of purulent wounds or skin ulcers by washing the wounds with the infusion obtained twice a day for 10 minutes.

WALNUT LEAF POWDER

Grind the walnut leaves as finely as possible and store them in dark jars in a dark, cool place. They should be kept hermetically sealed for a maximum of 2 weeks because the substances will evaporate quickly.

One teaspoon should be administered 3–4 times a day, on an empty stomach. The leaves have a strong hypoglycaemic action, are diuretic, and help the metabolism of sugars and fats.

It is a good adjunct for diabetes. It is recommended to take a teaspoon of walnut leaves 4 times a day for 3 months, and then take a month's break.

In cases of strong sweating, take a walnut fruit treatment 3–4 times a day for a month.

THE DECOCTION OF
THE WALNUT SHAFT

Ingredients: 300 ml water
Walnut leaves   40 g of honey

1. Crush the walnut leaves till you have 3–4 teaspoons of vegetable product.

2. Boil the crushed leaves with 300 ml of water for 10–15 minutes.

3. Filter and sweeten with honey, then consume it as hot as possible.

This is helpful to combat chronic coughs. The recommendation is to drink 2–3 cups a day to help with ailments. The taste is not very pleasant, but it is very astringent because of the tannins.

The tea has an effect by healing the tissues damaged by the infection, promoting the elimination of secretions from the airways, and reducing the sensitivity of the nerve endings that trigger the cough.

> **Walnut leaves absorb heavy metals from the atmosphere and purify the air.**

In some countries (such as Afghanistan and Pakistan), walnut leaves and bark are used to brush the teeth. They are known as *Dandas* in Pakistan. The use is common, as they have antiseptic and anti-bacterial properties.[16]

Other Uses of Walnut Leaves

- Walnut leaf tea combined with local baths helps to cure vaginal discharge (leucorrhoea). The general baths made

16 K. Vahdati – *Traditions and Folks for Walnut Growing around the Silk Road,* 2014.

with walnut leaves strengthen the body and help to cure various skin diseases, such as eczema and ulcers.

- In the form of compresses, they are used to treat eye diseases.
- For rich hair, you can use a compress of 30 g of dried and crushed leaves, boiled for 10 minutes in a litre of water, with which the scalp is rubbed every day for two to three weeks.
- Due to the pigment they contain, walnut leaves can be used as hair dye; the dyed hair has a very pleasant copper shade, keeping the shine without drying out the hair.
- In the treatment against sweaty palms and feet, you can use a tablespoon of dried walnut leaves, boiled in 200 ml of water until it reduces to half. Add a tablespoon of honey to 100 ml of tea. For 3 weeks, drink 3 cups of tea a day. Do not reheat the tea.
- Keep the tea in a dark bottle for later use, but only for external use. Dip a clean cloth in the tea and apply compresses to the palms and feet. This will remove any dry skin and moisturise your skin.
- Green walnut leaves have an insecticidal role; they can be put inside a wardrobe to keep moths, flies, ants and other insects away.
- Green leaves are placed in mattresses, dog cages, or hung in rooms in order to keep insects away.

In some of the Asian countries, the leaves are used for cleaning carpets and for making an ointment for burns, and they and are mixed with henna (= Kina). Nevertheless, in some regions, the leaves are used in baths to treat human body fungi.[17]

17   K. Vahdati – *Traditions and Folks for Walnut Growing around the Silk Road*, 2014.

Be Very Careful When Using Homeopathic Preparations!

The active substances contained in walnut leaves have astringent effects. Therefore, these preparations are not suitable for people who are underweight, those who drink a lot of coffee, or those who do not consume enough fluids.

The drugs are administered with caution in cases of hyperthyroidism, irritable bowel syndrome and acid gastritis.

In insulin-dependent diabetes, blood glucose should be closely monitored during treatment with walnuts; insulin doses should be adjusted to prevent hypoglycaemia.

Walnuts in their shells

Chapter 7

WALNUT WOOD

T HE WALNUT tree is a tall tree, reaching up to 30 metres in height. It has a broad, spreading crown, with numerous branches extending at right angles. The root system is strong and stretches over a radius of about 20 metres. At the age of 80, the main root reaches a depth of 5–7 metres, the lateral roots are 12 metres. The root system does not overcrowd, but after the top part of the root is dead, the offspring appear from the collar of the root. The trunk of the tree is straight and up to 2 metres in diameter. The bark is light grey and cracked.

Table made with walnut wood

Walnut wood is exceptionally and valuably useful if it is properly processed and cut at the right time. It is a heavy wood, very well textured, and has a very dense essence. New technology allows the walnut wood to be cut into the most durable and spectacular pieces, with a grand colour pattern in the interior and a great shape.

The former Romanian communist dictator Nicolae Ceaușescu used walnut wood to decorate his house walls and furniture. This and subsequent photographs in this chapter are from the museum Casa Ceaușescu (Ceaușescu Mansion).

It should be noted that the wood resists shrinkage deformations and can be bent and finished in many ways. It accepts a wide and varied range of varnishes and paints and can lend itself perfectly to the protective layers of varnish or oil, amazingly highlighting its spectacular fibre.

Wall covered in walnut wood

The interior colour of walnut wood varies depending on the variety. It can go from shades of light cream to a dark chocolate shade of brown.

Walnut wood is precious, being very durable and expensive. It is used in making luxury ornaments for boats, cars, interior decorations, jewellery boxes and weapon frames. Heavy, fine walnut wood is also used in the aeronautical industry, but mostly in the furniture industry in the manufacturing of solid furniture or panelling. In construction, it is used as a resistance element. Manufacturers of musical instruments or sculptors rightly consider it an ideal raw material.

Left: This door is made from walnut tree wood.
Right: Detail from a wall covered with walnut wood.

Chapter 8

CONCLUSION

T HE WALNUT would be called the tree of life, but we must not forget that all its positive qualities can turn against us if we do not associate its use with nature. There are many diseases where this fruit, whether prepared from leaves, or flowers, is contraindicated, and we must consider the requirements of our body before offering a remedy. If we are not sure about a remedy, then we need to read more, research more, ask questions and seek advice from specialists.

We must not forget that the 'royal' fruit is a large enough allergen, and those who suffer from allergies should eat moderately or not consume this product. Those with a slight allergy can still experience side effects such as rash, redness and itching.

The walnut is mentioned in the 'Song of Solomon' in the Old Testament of the Bible as part of the king's orchard. The tree is a notable presence in the Holy Land.

Although it is beneficial for the body, walnut consumption should be moderated or not used at all by people who have various skin conditions at certain stages (broken skin, advanced stage of psoriasis, etc.). In certain types of eczema or psoriasis, walnut kernels do not improve the disease if consumed in excess or if they are associated with other components.

Many people of the world attribute a magical status to the walnut tree, sometimes seeing it as a refuge for demons. In Romanian mythology, it is said that whoever plants a walnut will die when the tree reaches the thickness of its neck.

It is recommended to buy nuts, and less often walnut kernels.

The peel protects the beneficial properties of the walnut kernel. Let's not forget that light and oxygen gradually reduce the beneficial functions of the walnut kernel to zero if their exposure has been prolonged. In the Talmud, it is written that each walnut branch has nine fruits, and under each leaf stands a devil.

Keep the nuts away from direct light and humidity. Keep them in dark paper bags away from the cooker or window. If you have a pantry or drawer that you use less often, then you can put nuts inside. The walnut kernels can be stored in the refrigerator

Young walnut tree

or freezer in bags from which as much air as possible has been removed.

Frozen nuts can be consumed even after six to seven months. Before you eat them, you need to let the nuts rest for 15–20 minutes at room temperature.

In the arboreal zodiac, the natives of the walnut zodiac are born between April 21 and April 30 and October 24 and November 11, respectively, and correspond to the zodiac signs Taurus and Scorpio. Those born at these times of year are considered harsh, full of contrasts, and sometimes difficult to understand. They tend to aggressively defend their interests. Having a comprehensive mind, such a person is spontaneous and surprising in his reactions, but they are not flexible.

Although they are rigid, they are very ambitious and can be a difficult partner, given that they have their own conceptions, from which it is difficult to get out. If, however, you reach a compromise, you will find in this individual a skilled strategist, a person who is persevering and willing to go to the end to achieve everything they have set out to do. Although admired for their qualities, the native of the walnut zodiac is far from being loved by those around them. Even so, natives are beautiful, refined people, even if they are not very natural or well-mannered.

Talented and dangerous, they can be cruel and can make everyone cry, just because they likes to cry. It's different when it comes to being a life partner. Although they fall in love hard and very rarely, they will be a faithful partner for life, and they are willing to endure much in the name of love. They are passionate, but prone to jealousy and full of surprises. However, they will keep their partner in their shadow, which can often be danger-ous, and they can dominate their partner. Lunatic and a little sadistic, they do not back down from anything to ensure their success; they do not accept compromises; and they prefer to walk off the beaten track. Close, supportive people can easily have friends but especially many enemies.

Even though the only walnut variety in the world with

heart-shaped fruits grows in Japan, we can enjoy the walnut kernel, flowers, fruits and leaves of any walnut variety next to you. Open your arms wide and give a hug to the tree, or any tree you want, and you will be filled with the energy, health and peace of nature.

Dry walnuts inside their shell.

Nature peace.
Nature health.

REFERENCES

Balascuta Nicolae – *Protectia plantelor de gradina cu deosebire prin mijloace naturale*, Editura Tipocart, Brasov, 1993.

Beceanu Dumitru – *Tehnologia produselor horticole* vol. I, Editura Pim, Iaşi, 2002.

Beceanu Dumitru –*Fructe, legume si flori*, Editura Mast, Bucureşti

Bojor Ovidiu, Alexan Mircea – *Plantele medicinal şi aromatice de la A la Z*, Editura Reccop, Bucureşti – 1982.

Cociu Vasile – *Culturi nucifere*, Editura Ceres, 2006

Fischer Eugen – *Dicţionarul plantelor medicinale*, Editura Gemma Press, Bucureşti.

Gradinariu G., Iastrate M. – *Pomicultură general şi specială*, Editura Tipomoldova, Bucuresti, 2004.

Georgescu T. – *Dăunătorii pomilor şi arbuştilor fructiferi prevenire şi combatere*, Editura Ion Ionescu de la Brad, Iaşi. – 2004

Josan Elena – *Comoara sanatatii*, Editura Artmed, – 2005.

Lawrence Sandra – *Witch's Garden*, Editura Welbeck, Dubai, 2020.

Mihăilescu Grigore – *Pomicultura Ecologică*, Editura Ceres, Bucureşti, 1998

Messegue Maurice – *Alune, nuci, migdale*, Editura *Venus*, Bucureşti, 1998

Negrilă A. – *Sfaturi pentru cultura pomilor*, Editura Agro-Silvică, Bucureşti, 1961

Pârvu Constantin – *Universul plantelor*, Editura Asab, 2006

Pleşa Maria-Marta – *Dulciuri pentru toate anotimpurile* – vol. I, Editura Ceres, 1990.

Powell-Hulbert Charles – *The walnut tree, tales of growing and uses,* Unicorn Press, 2019.

Ungureanu Ion – *Circulaţia sevei brute şi rolul ei în creşterea şi rodirea pomilor fructiferi,* Editura Conphys, Râmnicu Vâlcea, 2008

Vahdati K. – *Traditions and Folks for Walnut Growing around the Silk Road* – (Department of Horticulture / University of Tehran / Aburaihan Campus / Pakdasht, Tehran), Iran – 2014.

Woolf Jo – *Britain's Trees,* Editura Pavilion, London – 2020.

*** – *Conservarea fructelor pentru iarnă,* Editura Ceres, Bucureşti, 1978.

*** – *Dicţionarul plantelor de leac,* Editura Călin, Bucureşti, 2008.

*** – *Plantele medicinale,* Editura Alex-Alex, 2002.

Online:

www.fai.ro

www.sanovita.ro

cookpad.com/ro

www.revistadinlemn.ro/2016/10/27/nuc

enciclopedia.asm.md/?p=6786

ro.wikipedia.org/wiki/Nuc

ecology.md/md/page/12-curiozitatsi-despre-nuci

www.nucicultura.ro

www.nucifereregia.ro/polenizatorii-din-livada-de-nuc/

sfaturipomicole.tripod.com/id11.html

www.researchgate.net